D1289996

Lit
851.912
L979
i

7477777

DISCARDED BY
MEMPHIS PUBLIC LIBRARY

MAIN LIBRARY

Memphis and Shelby
County Public Library and
Information Center

For the Residents
of
Memphis and Shelby County

In
the Dark Body
of
Metamorphosis

& Other Poems

Books by I. L. Salomon

POETRY
Unit & Universe (1959, 1961, 1963, 1964)

POETRY IN TRANSLATION
Carlo Betocchi: *Poems* (1964)
Dino Campana: *Orphic Songs* (1968)
Alfredo de Palchi: *Sessions with My Analyst* (1971)
Mario Luzi: *In the Dark Body of Metamorphosis* (1975)

In
the Dark Body
of
Metamorphosis

& Other Poems

✦ ✦ ✦

MARIO LUZI

Translated by I. L. Salomon

W · W · NORTON & COMPANY · INC ·

NEW YORK

Copyright © 1975, 1974, 1973, 1972 by I. L. Salomon. All rights reserved.
Published simultaneously in Canada by George J. McLeod Limited, Toronto.
These translations of the poems of Mario Luzi are printed with the permission of Giulio Einaudi Editore, Turin; Aldo Garzanti Editore, Milan; and Rizzoli Editore, Milan and New York. Printed in the United States of America.

FIRST EDITION

◆◆◆ THE TEXT of this book was typeset and composed by the Vail-Ballou
◆◆◆ Press, Inc. in Linotype Janson. Printing and binding were done by the
Vail-Ballou Press, Inc.

Library of Congress Cataloging in Publication Data
Luzi, Mario.
    In the dark body of metamorphosis.
    I. Salomon, Isidore Lawrence, 1899–   tr.
II. Title.
PQ4827.U915   851'.9'12   74–11122
    ISBN 0-393-04391-6
    ISBN 0-393-04403-3 (pbk.)

1 2 3 4 5 6 7 8 9 0

7477777
MEMPHIS PUBLIC LIBRARY AND INFORMATION CENTER
SHELBY COUNTY LIBRARIES

T O

Adam
Beth
Martha
Tamara

Contents

Acknowledgments

I owe a great deal to Professor Rolando Anzilotti (University of Pisa) and to Professor Sergio Baldi (University of Florence) for making superlative comments during my work-in-progress. To Miss Maria Gargotta, librarian at the Italian Cultural Institute, to Professor Tom O'Neill of Trinity College, Dublin, and to the man of letters, Sidney Alexander, my deepest thanks.

My thanks also to the editors of the following periodicals for first publishing a number of these poems, a few in earlier versions: *Mundus Artium*, the *Arlington Quarterly*, the *Massachusetts Review*, *Forum Italicum*, the *Michigan Quarterly Review*, the *Mediterranean Review*, *New Directions 27*, the *Christian Science Monitor*, the *American Pen, Nimrod, Translation '74*, and my immeasurable gratitude to the National Endowment for the Humanities for a grant giving me time to work.

Introduction

Mario Luzi is the foremost Italian poet of his generation. He ranks first as an interpreter of life. He owes much to ancestral ties and to the Tuscan landscape of his youth and middle years. He is at home with such rational philosophers as Bertrand Russell and Adorno (Theodor Wiesengrund). Essentially, he is an apolitical humanist.

Indebted to Dino Campana, slightly influenced by T. S. Eliot, Mario Luzi has been prodigious. He has published more than a dozen books: of poetry, essays on the craft, and translations of *Richard II, The Rime of the Ancient Mariner,* the poetry of Mallarmé and Guillén. He writes a weekly literary column for the *Corriere della sera,* the leading newspaper in Milan, and contributes frequently to the major Italian quarterlies and periodicals. He is also professor of French literature at the University of Florence.

He was born in Florence on October 20, 1914. When he was twelve his father, a State Railway employee, moved the family to Siena, where the boy entered the *liceo,* the academic high school, in preparation for the university. He studied Greek and Latin, the mainstays of a classical education; a natural linguist, he also became proficient in French and Spanish, and acquired a reading knowledge of English. However, he did not do well in studies that lead to academic success. Unlike a young scientist, who knows where he is headed and of necessity learns the ABC's of his calling, a young poet knows only that he is in love with the sounds of words, with poems or phrases in poems he has learned by heart. Indeed, the apprentice poet often simply does not have the skill to create a fully realized poem. An awareness of his predecessors doesn't immediately help. He may know Homer in Greek, Virgil in Latin, his countrymen Dante and, more recently, Foscolo, Leopardi, Carducci, D'Annunzio, but

11

these towering influences overwhelm and enslave him. The young poet must free himself from their bondage.

As a poet, Luzi first belonged to the "hermetic" school that grew out of the suppression of thought under Mussolini. Confronted by totalitarianism, writers were constrained to conceal ideas to which the state was antagonistic. Francesco Flora, a critic of the time, named the movement *ermetismo* and characterized the style as cryptic, difficult, obscure.

Italy had passed through a social upheaval on Mussolini's coming to power in 1922. The *ermetici* had sealed themselves off from politics; Ungaretti, in his forties, and the younger poets Montale and Quasimodo were writing in *uno stile nuovo*, a new style, intended to bury the cadaver of the immediate poetic past. And Mario Luzi, all of eighteen, a senior at the *liceo*, wrote a brief lyric unaware that he was following the principles of Imagism laid down a generation earlier in a manifesto by Ezra Pound, who was now living as an expatriate in Rapallo. "Just a Touch" is pared of verbiage. There is not even a verb.

As a freshman at the University of Florence, Luzi saw his first poem in print in a "little" magazine. He was soon to appear in such formidable periodicals as *The Frontispiece* and *Campo di Marte*, named after the proving grounds near the Florentine armory. After taking his degree, Luzi, in order to make a living, taught French in the high schools of Parma, San Miniato, and Rome. He was absorbed in French literature and explored and interpreted the aesthetic theories of the classicists and the modernists (innovators and experimentalists) in poetry.

Luzi wrote fluently and prolifically. His work was admired by his elders and hailed by his peers; his books were much sought after. In 1960, he was invited to compile for publication all the poems he had written through his fortieth year. These collected poems, published as *Il Giusto della vita* [*Proper Perspectives*], mark the close of his hermetic period.

During the late 1950s, Luzi freed himself from the strictures that had bound him to the hermetic school, and developed a new technique, writing lyrics more mature in concept and execution. He let these poems weather for five years. In the meantime, he published two books of essays, *The Napoleonic Generation*

(1956) and *The Symbolist Idea* (1959). He wrote a series of dramatic quasi-dialogues, *In the Magma* (1963), long-line poems in a freer style. The speakers are of his contemporary world, old friends confronting him with the desperation that turned them politically left in opposition to Italian fascism, only recently defeated and not wholly discredited. In the contrapuntal give-and-take of dialogue, Luzi takes his stand:

> I join with every unstable power
> of resentment and unfulfillment; I touch
> the unsociable element that keeps the world
> hanging.

After a long seasoning, his lyrics written between 1956 and 1960 appeared. He entitled this book *From the Heart of the Countryside* (1965). In "The Fortress," he speaks of a man, "gripped in the vise/of old age," and observes:

> Life, if it lasts, is no more than the slow
> extrication of souls, one from another
> at the exit of the maze . . .

In "Shots," Luzi, who has never shouldered a rifle, says of the hunter:

> so man trembles
> and as he meets his suffering head on,
> he writhes between cowardice and courage.

And speaking of himself as a child in "How Much Life!" he is somber as he recalls:

> No one ever perceives of life
> as strong as in its doom.

On November 4, 1966, a Noachian deluge almost submerged Florence. Not for a hundred years had the city undergone such a cataclysm. The great museums, the great libraries, the great churches of that crucible of Western culture, among them Santa Croce with its tombs of Galileo and Michelangelo, were flooded. Countless books and innumerable illuminated manuscripts were soaked by slimy water and caked with mud. At the first subsiding

13

of the Arno, it was the hippies in bleached jeans who voluntarily repaid their Florentine hosts and saved what they could of paintings, books, and manuscripts.

Those days were grievous enough for Luzi, and they led him to write two long poems which rank as masterpieces: "In the Dark Body of Metamorphosis" and "The Whirlpool of Sickness and Health." "Dark Body" first appeared in a 1969 issue of a periodical, *The Literary Landing-Place;* it is the most important poem to come out of Italy since Campana's "Genoa" appeared in *Orphic Songs* on the eve of World War I.

"Dark Body" is a poem in seven sections, each with episodes consistent with its theme. It begins with a philosophic abstraction and moves swiftly to the drowned city. Luzi weighs the culture of his age against the cultures of the past. His pilgrimage is the real and symbolic proof that man survives disaster and hope is ever with him.

One morning in the late fall of 1970, Luzi brought to my hotel room in Florence the poems of "The Whirlpool of Sickness and Health." Their brilliance leaped from the typescripts, which were not copies of the manuscript already in the hands of his publisher but an unevenly typed sheaf transcribed from notes on scratch paper. That afternoon as I worked at translating them, I wondered how they compared with the originals. I was troubled until the galley proofs arrived a month later and proved that Luzi's memory had not betrayed him.

Mornings we discussed the several versions of a single poem I had translated the afternoon before. When my Italian failed me, he gave the closest approximations in impeccable French. "The Whirlpool" revealed a deep introspection that reflected the image of his private world.

Besides these renderings into English, Mario Luzi's work has been translated into French, Greek, Romanian, and Russian; at present it is being translated into Swedish. My translations are my tribute to him. Truly a great poet, he will endure.

I. L. SALOMON

New York

14

IN THE DARK BODY

OF METAMORPHOSIS

Translation dedicated to Allen Tate

In the Dark Body
of Metamorphosis

. . . because they are such, the events of epochs
occur in them . . .
                    ST. AUGUSTINE

I

"Life consonant with thought abstracts us from the
        sources of thought,
life consonant with life
leads us into error and suffering where life is impossible"
the wall of a dream dreamt wide-awake
thrusts me back. "Impossible
to live, even to think"—a cracked rock
bears this inscription; take a good look:
a cobweb of wrinkles, the defeated face
of a master-thinker of the western world
where nothing lives but two points—two eyes
        of a wren—and silence.

The world is greater than that—I smile
and I think of my hilarity as a covey
in flight from a house collapsing.
Lose yourself if you wish to find yourself; desire
so as not to have—lightning
slants across me and brightens my vision
perhaps from my side of innocence
that like water has withstood the millstone
and for this, this, does not surrender.

"Pray," it says "for the drowned city."
Coming to me face to face from the past
or from the future, an apparition hidden
behind a flashlight searches for me
in the oozy mire of the deserted street.
"Quiet," I plead, uncertain it may be my soul
returning to my body lost in that mud.

"You who have watched until sunset
the death of a city, its last
furious pawing of air in drowning,
listen now to its silence. And wake up,"
that wandering apparition continues,
I not quite certain it may be another's, not my soul
in search of me in the ominous marsh.
"Awake, this silence is not
the mental silence of a profound metaphor
as you contemplate history. But a brutal
cessation of sound. Death. Death and no more."

"There is no death that is not also birth.
Only for this will I pray."
Injured, sloshing in slime, I speak to it
while its light, eclipsed in an alley, flashes,
and its after-image sends a glaring reflection
ambiguous and visible to the lynx and the mole.

How many lives, this for example
called my own out of inertia and habit . . .
And now with a bewildered glance
a figure surfaces,
ripping a membrane of rain
from the depth of the fish-swarmed city,
takes her son by the hand, a hand
it seems to me, escaping her grip;
gasping, she says not a word

while I feel the pain
beyond that cause, and waves
of remorse strain to a spasm
the infinitesimal instant of time
where the event is suspended or the pulsar.

2

"To walk about to the chattering of a bazaar-hawker,
to sell what's worthwhile or even perpetrate a swindle,
to bargain and plot in self-interest,
aptly wearing black for the lost revolution . . .
well, worse has been destined under the sun."
I used to joke with my one-time friends;
aware and unaware that with fire I aggravated
a malign wound spurting pus.
"Vicious" just as I came into the gleam of their glasses
those watery cetaceous eyes
equally sensitive to power, good business, remorse
scalded me alive.

The leafy bough seen once again
bare as a twig,
that place of our summer excursions
come on out of season, bleak,
bony, nipped by cold—in the meantime I was thinking.

O my youth! for man lost
in limitless love
without return of conscience, the point
between memory and desire shifts,
adrift in a whirlpool.
Time past and future reverse direction
and capsize; dolphin
and tuna in the net of perception.
It's I who am on the wrong side, amen.
Except for the gift of speech. Less

the Pentecost of grief
that smelts everyone in the same mold.

And that wine. That wine not dulled in the throat.

❖ ❖ ❖

The shadow between lovers of unequal ardor
is to her detriment, who suffers a little
and slightly amused and wise,
smiles, divorce behind her
not altogether bitter.
Only slightly slackened is the thread
of many good habits, trips, the arts,
a dignified Christmas at Zermatt.

Of all this she excuses herself with humility
and grace, yet frivolous only so far,
time for a greeting or rather a glance.
She excuses herself as I hardly look at her
and into her thoughts I plunge one of my own,
pleasure-minded in this light on the Lungarno.

❖ ❖ ❖

"Sufferings that come,
that go, besmirch you.
In the interval they mature you, carry you to a point . . ."
The voice of a woman always heard,
once my mother's, now hers, the sacrificial
voice that dissolves the knot
loving and grievous in every life, severs itself
from some exchange of words,
opaque in their meanings in the cavern of the year,
not in the spring but in the haze of its birth.

Voice deprived of speech in her throat
who temporarily marked
her pain there

and shut it in the strait
of fear and anxiety
during a spat in the kitchen, a pleading on the stairs.
Impersonal and steady, wearied by the sea-change
the voice bores through, recasts every substance
        from the beginning,
the city in stone, history by its events.

"You who boast of familiarity with the sea and have it not"
a useless gruff cry, evocative of bodies bound to the mast,
ears stopped with wax, warns me
not to ignore its sweetness, not to betray remembrance
but to go my way, do what I have to. And may it be right.

3

Across the troubled sea
of washed clays,
become a downy green in March,
the crooked road from Siena that goes on to Orcia
is a road outside of time, an open road
pointing as it winds to the heart of the enigma.

Real or unreal, solar or nocturnal—
my ancestors from father to son
with foreboding torment
attentively followed its ups and downs.
Real or unreal, solar or nocturnal—
the mind interrogates
the years (and the idea of life is speckled
in a face doubly unassailable),
examines the rough plant on the moor,
the burning hillocks, the scattered rocks,
and the wind, I know not whether of time or space,
        that thwarts the blood.

Thoughts drawn on the string
of endless questioning

do not permit existence, provide no answers.
She indeed understands this who walked through those dunes.

❖ ❖ ❖

"Not to distinguish, not to divide. Take
the good as it is offered you."
He who goes forward from the brighter side
wants to reach me, hammers blows on the tunnel's diaphragm,
a solar alter ego of the event, my joyous counterpart.
"Preserve wisdom for later, for another time."
I still resist him, resist him as I can.

Now we are not far from Tiflis
between Asia and the West in that hour
that splatters ink from a skyblue speller on the mountains
when he resumes: "For love of shade?
Of the world's and your mind's rage?"
"Not only this," I grumble under that lash
in the air broken by a shuddering between the Caucasus
        and the Caspian.

❖ ❖ ❖

She who thinks of parks in autumn
has in her eyes and hair
something of the tenderness of trees;
as if behind a silent waterfall,
he grows old before her,
who is aware of my conjecture
as in the rolling train I observe her;
she does not find my smile indiscreet
but welcomes it in an exchange
refracted unendingly in her own.

4

But there is an even greater inexpressible sense
as into the empty molds of history

a new metal flows, barely adequate,
nor are others found ready for the casting now,
this irreversible instant, or never.
And in that dispersion of power
sick in the will or drugged
a clot still called soul spins
about by itself and breaks up, its movement impeded:
not by a little but by too much ardor is it consumed.

Or rather isn't it a message like radar imperceptible
        to the ear that worries you,
while you observe (it is not rare)
its world vibrate in sparkling vertebrae
in feverish cartilage—I ask myself.
And at that moment someone screams
in a distorted voice: "Come to me,"
one inured to the sea, in out of the rain,
one would say, with two hailstones or two white holes
        for eyes,
ordering me to be cold-blooded and calm,
piloting me outside a whirlpool
swarming with slag—or at least he'd like to,
while I between deference and reluctance
thrust forward yet withhold myself
in the sluggish air of my room still blue with night.

Or when time under pressure
scatters its futile power
in a whirling cloud of slag
and your self in one part of you—you know not
what precisely—suffers as you would sleep,
a restless
semiconsciousness keeps you awake
not completely present at the metamorphosis
and the long grief of the birth of an epoch.

A sense of strength

dissipated by the world, by disunity,
derives from it and sickens you.

"But how small a thing your lament"
a childish glance, a reproach
in the budding depth of the eyes—the deep blue
glance of creation, it seems to you—
which pierces your retina with its throbbing and brightens,
driving you from the enclosure
of the soul's infirmity, calling you to the future
of an expanding universe. This calls for no answer.

"Didn't you know or remember
torments like this of times
more and less mature than you?"
The soul half-awake hesitates
or rather the night itself
torn by the blinking of chemical hues—insomnia.

"Didn't you know or remember?"
A ruinous wind sensed by proconsuls
broods over and over
in some dismantled province.
Didn't you know or remember?
The ricefields of Vietnam gurgle from under the film
         of muddy water.
"The memory of this is worth little, little."

In this hotel
in this hotel—they remind me—Esenin killed himself.

5

Church, Church . . .
GIOVANNA MARINI

Something imminent dominates her,
makes her weep and offer me tears in silence
while I not circumscribed by any limit
of the past and present look at her
and do not disturb the silent session
with demands or anything else. I study
her eyes lowered to and lost in the arabesque design.
And I receive the power of love and sorrow
of the world. And still more, still more than this.

"I almost don't remember him"—she means her first husband.
"He no longer had a face,
since man's was destroyed
and perhaps still is, granted the concentration camps . . ."
The sound of metal hammered cold has a timbre
that glances off me not very high in the scale
at a point between animal and man
who knows whether ever reached by the swirling fire of
        redemption,
surely not ever risen to the wind and light,
not for this dead or lost. Grief
that follows dulled, diffused
between the liane knots of mankind,
not mine, not at heart's height
not of the horizontal arms of the Cross—

                                    what do you know
who listen not to her but to a manikin emptied of memory that
        resembles her a little,

squeaking in the washed-out air of the loss of man with no
        contours to features
in this aftertime, in this sickness of non-love that overflows—

rather stretch your arms to her
while in the blue inner light
springtime drowned in the grass
strikes her ear-drums, blood pounding,
beats its rhythm on a triumphal, if humiliating, drum.

The city empty the holiday afternoon
a scull threading its way from bridge to bridge
on the shadowy river in search of a rhythm
while she who is like sandalwood
perfumes the axe that cuts her;
she reopens the house replete with loneliness
but almost with a gracious smile;
at the moment I know it
when her felicity photographed by a flashbulb
reappears in front of her.
I don't call her back. I don't detain her.
Not this thought as if thought by another, or not even that,
wrenches her from mutation.

A lamb, a victim of troubled awakening,
she sits now in her corner
crushed—it may be—
but erect in her bitter dignity that remains of understanding
she spends heavy futile hours
listening with different ears to music once heard,
releafing books read with a mind transformed,
and she withers and glows in renewed solitude,
strange event, even majestic, that befalls her
no grander than herself who welcomes it between soul
        and the life-source.

—Do not think of yourself as guilty or innocent.
This is not the point: if you
more than others happened to offend her.—
Meanwhile her shepherd smiles,
the Shepherd of her anguish, John,
suspended in that seismic breath and fixed no less
in the light of the celestial quarry that surrounds her:
while I perceive the doing,
the undoing, the continual source, the hive.

6

The obverse of felicity in an epoch:
the queen of the rocky city,
bright of mind, her weeping
and the humbler tears of her subjects
in no way caught up in the perfection of work—

yes, the solar immutability of number.
But, before and after, the fear of change,
its necessity. And the spirit sick at a stage where
        it not only has no peace
but wants no peace, desires nothing,
rejects nourishment, rejects life.

Of this I think as I loiter on the scorching stairs
        among a few others astonished
as a pigeon whirls a blue oil of dizziness
from one slope to another in noon light
in the miraculous square that has no true shadow
but a thin crack here and there in the marble walls.

An instant that holds and dazzles
marked by a flying line of lofty cupolas
sheltered, it seems from metamorphosis—
the consummation of power and art:
But the other, the unassailable part of the fire—I tell myself
and think of the long antelope glance of local women.

Nevertheless: "Follow me."
Life, its skylark warbling piercing the meshes
of the Sunday fusillade,
transfixes me asleep
while I bound to the water-wheel
of change in the world
(also, I tell myself, with the wings of a hippogriff)
smile and don't answer.
"Follow me" that scream repeats,
already more distant as if half-spoken by a harp.

I confide it to you who already read my thoughts
and I feel no shame, not even torment.

The sudden blossoming of the very soul
in the full morning sun of absolute faith
shared with me, rather unique
and possibly universal—is this
that darkly awaits, I am certain,
the fixed magnet of the mind,
fixed in its great transits
when love is born—who knows—
or Trotsky's train breaks through
the tough perspective of history.

Hope—I know little about it.
Or else it blazes in her face
so illuminated I am reminded of
a cloud of fire in a stand of oaks
a little above the snowfield. Without this,
I say, even less. I'd know even less.

7

The living instant, the springtime of the world
that blazes and recedes into infinity

in another's eyes
now when thought shared
completely suppresses shadow
and the spoken and the still unutterable
sparkle in identical minds,
the living point, the budding root of her continual
     beginning—

she freed herself from the past,
cut across my sleepwalker's path
a little like a bird
surprised thirsting in flight down hill
and came to meet me face to face on the brink
where I walked in danger
looking for herbs—
remedies for the ignorant.

And no longer can she be the same
or have another replace her
to perpetuate and overwhelm her
and make her cry . . .—I think
years—or ages later. As I look into her eyes
I come on the mutable and the eternal
intimately mixed in the source.

PROPER

PERSPECTIVES

Translations dedicated to Rolando Anzilotti

Just a Touch

Suddenly April: watery and dusky
skies an irritation;
the quietness of a mat
on a window, a touch
of wind, a wound;
this life, an alien presence
in doorways,
in streams slightly ashen,
in your footsteps, an echo under the arches.

To Spring

Ships submerged in the seas become a grassy trail
for the fledgling swallow crossing the continents.
Seamen on the windless ocean
mirror hardened faces
and their ephemeral years in the flux of eternal water.
Fear for and humility of life,
ardent convictions, sink to the blossoming earth.
In churches mothers surprised at themselves
and untiring truths wipe away their sadness;
they remain under the burning lamps with their childhood.

Springtime for Orphans

Spirit of green watersheds
the sky moves with the motion
of the sea where waves
and colorless sails wobble,
turn the Virgin's eyes
to the hearts of lonely children
and spread a skyblue mantle
on their nakedness.

In the eyes of the children
raw poverty glistens and love silent
these many years in the breast and the immeasurable
sadness of denial
of dreaming on and on. Above and unknown
a motherly face throbs for them
in a golden radiant smile

and it is present during chilly vigils
without fire or speech
where time in old desperation
runs toward a dark outlet; the blood
totally hurts and stings
in the smarting dismay of a greeting
to the lost mood, loneliness.

Natural World

The sea is attuned to the land
and everywhere above, a more joyous sea
for the swift flash of sparrows
and the track
of the restful moon, the sleep
of sweet bodies ajar to life
and death in a field;
for those voices that come down
eluding mysterious doors to leap
over us as birds, singing, crave
to return to their native islands:
here someone prepares
a purple couch and a song that cradles
anyone unable to sleep—
so, the stone was hard;
so, love poignant.

Ivory

The ever-dark cypress is alive,
the somber mountain buck is elated;
in reddened springs the mares
slowly wash caresses out of their manes.
Down from misty forests immense
rivers lash against the towering cities
constantly; quivering sails move
in a dream towards Olympia.
Airy girls will travel the crowded roads
of the Orient and from brackish markets
will look cheerfully at the world.
But where will I draw my life from
now that flickering love is dead?
Roses corrupted the horizons;
faltering cities sprinkled with troubled
gardens remained in the sky;
her voice on the air was a desert rock
never to be heaped with flowers.

Black Flowers

Your lifeless hands that darkness tends to
and silence leads astray
already gather the black flowers of Hades,
clammy flowers thick with rime.

The meadow-saffron more touching than your smile
that fever consumes
bows to the dim fields of Elysium,
fields dulled numb in the depth of winter.

In the wind your drowsy body, a lonely star,
is radiant between window-panes
and your muffled footstep is no more than
the lateness of roses in the air.

Neither Time

Your afflicted hands grip the bank,
you climb up, enter the district,
take the somber road of the dead.
I do not know your name. Perhaps it is Acconsolata
or Apparita or another among the countless names
whose meanings were so long concealed from me,
not their portent when given in homage
to a wayside shrine or to a bend
in those roads that spell exile.

This is our boundless region.
Gather its forlorn flowers, dull grass,
a crop that wavers untouched; rest.
And I don't know what that bleached grass is
in the abandoned field
where hope cannot survive.
Now mow the gray stalks, the dismal
veil of spikes lost in the distance.
Penetrate the unending thickness.

Here is the realm that we must pillage,
its abundance scythe, but neither time
nor craving prods us any longer.
This is our sunless region;
this, the endless color of pain
known to you only as an intimation
where a wall opened up or an apparition
in light among Lenten flowers . . .
and what we lost and suffered unknowingly
except in signs bears a name.

I recognize the desolate country
of our birth without a beginning,
of our death without an end.
It is this I had called chance,
adventure, fate or night
or those awesome names
my anguish told me,
not compassion that penetrates, sees.

Caught in Light

Caught in light, someone stirs between the walls . . .
perhaps it was you, now it's a ghost
or perhaps it's everything that has no peace,
place, motion and it is neither true
nor insubstantial—an empty thing that only
perfect mirrors reveal trembling.

An incorporeal image, never at rest . . .
it is ours. I thought it a chimera
when someone unconsoled appeared miraculously
under arid hillsides
on dark roads where nothing lives any longer,
nothing except hope for thunder.

Eternal Becoming

What birth, what death, what seasons!
Shadow that you are, pulverized on this jetty.
If a window shudders, the promise
of flowering sparkles and trembles in vases.

Birth and death, swift truths . . .
We are here as we must be at one place
in a point of time in a room,
in light, in eternal becoming.

I know of no other fate that is not this.
I sit, enraptured by this delicate flame;
I look at the clear febrile
metallic day as winter takes hold of the sky.

In the House of N,
Childhood Friend

During Lent the wind is harsh;
it moans through the cracks and under the doors;
hissing, it invades the rooms and rushes by;
outdoors, it tears the ribbons of paper streamers
to shreds; if some, caught
on the wires, sway and quiver,
it whips them, drags them whirling.

Here I am, a person in a room,
a man at the back of the house. I listen
to the sputtering flame, the louder
throbbing of my heart. I sit. I wait.
Where are you? Even a trace of you gone . . .
whether I look here at the fury or the grass
beyond, there is the impoverished gray of the mountains.

Village-Fate

Around the fire we talked at length of you
after evening devotions
in these gray houses where impassive time
bears and banishes the face of man.

Later the conversation turned to others and
        their possessions;
there were marriages, births, deaths,
the dismal ritual of life.
Someone, a stranger, came by and disappeared.

I, an old woman in this old house,
sew the past to the present; I weave
your infancy with that of your son,
who crosses the square with the swallows.

April—Love

The thought of death goes with me
between the walls of this road that rises,
its windings difficult. The springtime chill
troubles the colors,
vexes the grass and the wistaria,
roughens the pebbles, pinches cold hands
under mantles and raincoats and makes one shiver.

The season for hardship and pain, the season
that in a whirlwind whisks flowers brightly
mixed with harsh apparitions; while
you ask yourself what it is, each
vanishes swiftly into dust and wind.

The path is through known places
except that unreal signs
foretell exile and death.
What are you? What have I become
wandering in so windy a space,
a man following a light and weak trail!

It is unbelievable that I look for you in this
or any other place on earth
where we can hardly recognize each other.
But it is still an age, mine,
that expects from others
that which is in us or else does not exist.

Love helps us to live, to endure.
Love destroys yet gives a beginning. And when
he, who suffers or languishes, hopes (if he still hopes)

for relief to be sounded from afar,
there is in him breath enough to quicken it.
This I have learned and forgotten a thousand times;
now from you it returns to me, made clear,
now it takes on liveliness and truth.

My penalty is to last beyond this moment.

Meeting

It is not love but this unfamiliar road remaining
from me to you, from me to others
still tempts me. I come on
years at the foot of trees, years and fallen
berries and from the crossroads
a clutter of leaves
scampering or lifted in flight. Desires
and suffering crowd in the mixup
and I all but pass through and freeze.

                                      Time,
you say, fulfills its work,
tears the fleece from the avenues, lights
the pyre. I am become empty,
a shadow that varies its place in the flame
of perpetual death. And who are you,
a real person or a spirit,
who comes again in a dream at this turn?

                                    Look at me:
I remain after so many and so few years past.
I am transformed from a girl into a mother,
and a mother even overcome keeps faith,
holds firm to the earth or pretends to,
for her son must grasp life
and suckle that field even if withered.
This toil will never end.

The wind that deflects the ball from branch to briar
upsets a child's play,
scatters the embers; and you who spoke just now
are silent . . . a moment in our lives.

At last the sun gathers its rays
on the sky's threshold and slowly
moves off, and still the wind has no respite.
On the tree-tops where the dying red breath
of light lingers, a few leaves
swirl down to join the rest.
Nothing else: the hour says we must
each resume his way
in this tug of souls and remains.
You precede me but do not know if truly
on this night there is illumination.

Birds

The harsh voice of the wind is a warning
to us, a covey that at times finds peace
and refuge in these dry branches.
The flock resumes its grievous flight,
migrates into the heart of the mountains, purple
dug out of inexhaustible purple,
a bottomless quarry in space.
The flight is slow, breaks wearily
through the blue that opens beyond the blue
in time beyond time; a few birds
tumble sharp cries
with no wall for echo.
That which resembles us is the swaying of tree-tops
at the time (one can hardly think
or speak) when on invisible stems
there blossoms everywhere a novel springtime
in grazing clouds the wind
pastures in a damp or scorched sky,
and the day's lot is varied:
hail, rain, radiance.

Mortality

What do you expect who come from so far-off
and enter the fog flying blind
here where even birds from branch to branch
lose their trail home?

Life, as it must, preserves itself,
ramifies into a thousand rivulets. A mother
breaks bread for her little ones, charcoals
the fire; a day flies
or is irksome; a stranger arrives, leaves;
snow falls, it brightens, or a late
winter drizzle smothers colors,
soaks shoes and clothes; night comes.

It is little—there are signs of nothing else.

At a Point

When spring arrives
and the body still alienated
withstands the blow, it trembles
stirred to its roots:

or before, even before, in a night
of starts and anxieties when a dog whines
between the stars and the frost
and a wind close to the ground gathers wings
and strength from the year's grief; and by that howl
an animal, deep in its den, wakens;
the shepherd in seclusion lifts his head to the peaks.

I have no peace. I sigh for you,
my soul, and from known and unknown places
where you were trampled.
I tell you: hope (I wish it): be still.

At a point in the wind,
at a point in the eternal tempest,
in weakness or in cowardice, I set snares,
I prepare to deceive you,
I lie, someone will control
you; guides will come . . .

The poor man is in the dark about everything when
         he suffers,
suffers without courage, without precaution . . .

On the Shore

The deserted landings straddle the breakers,
even the old man of the sea is sullen.
What are you doing? Me, I trim the lamp.
I keep this room alive when I find myself
in the dark about you and those dear to you.

The gathering, scattered, comes together again.
After these storms we take a head-count:
You, where are you? I hope safe in some port . . .
The lighthouse keeper goes out in his skiff,
scans, patrols, heads for an opening.
Time and the sea take on these pauses.

Jeweled Light

Wind of autumn and passion. And dust,
dust that crawls over the land,
its roads blanched whiter than bone.
This, the moment, when the crushed heart throbs,
doubtfully annuls what was real,
not fable, not empty apparition.
What can I gather from news of you?
I know you well enough to recognize
your unease. I am certain you hardly dare,
if you really dare, to ask yourself what I think.
I think of you, of your love unfolding
under the jeweled light that is Umbria
in early summer between Foligno and Terni.
I ask myself, excuse my madness, whether
joy will be joy forever
no matter how full the measure is
of things that I must love and lose.

As You Will

The northwind cracks the clay,
presses, hardens the worked soil,
provokes water in cisterns, leaves
hoes fixed, plows idle
in the field. If anyone goes out for wood
or moves wearily or hesitates
benumbed in his cape and hood,
he clenches his teeth. What prevails in the room
is the silence of the mute testimony
of snow, rain, smoke,
the immobility of change.

I am here putting pine cones
on the fire. I listen
to the windows rattling. I am neither calm
nor anxious. You with your everlasting promise,
come and occupy this place
left by sufferance,
do not despair of me or yourself;
rummage around your house;
look for the gray shutters of the door.
Little by little the measure is full,
little by little, little by little,
as you will, my loneliness spills over.
Come in, enter, plunder.

It is a winter's day this year,
a day, a day in our lives.

Indoors

Sunny Sundays we penetrate the hidden valleys,
swarm through and return gratified
with flowers and branched clusters for vases
in corners or on the kneading-trough in windowlight.

In this open book of months and years,
I lose my place. I smile; I see,
if I raise my head, two bright windows
where there's the trembling expectation of swallows
and you, lifting these slight trophies.

One day. What day? Between this spring
and winter, one year among many
you and I, and between us our son:
from room to room this limpid light.

The Fisherman

Folks come down for water. Mute,
they brush past the bulwarks of ships at anchor;
they are startled by the jolts in docking.

A slight
early summer breeze blows softly, grazes
the tents, the grass, ruffles the hair.
It is dawn, the hour when nets are drawn in,
the hour that in a shudder of expectation
and luminous uncertainty lances
from house to house, creates blanks and images,
that if looked at closely, dissolve
quickly over the trees and beyond the bridges.

Time hangs on something somber
and obvious, when it seems certain
that truth is not in us, but in a secret
or a miracle nearest to revelation,
time that illudes men, and if it awakens
hope, it is the hope of a portent.

Disquietude makes the shadows
remote and eerie there at the water-edge
and on the wet sands I scrutinize
between these spars and these dwarf trees.

Forgive me! It is part of human nature
to search as I do in elusive places
for what's closest, humble and true to us,
or else nowhere. I crane my back.

My eyes anxiously watch the fisherman
as he appears on the breakwater and carries
from the sea what the sea lets him take,
few gifts, from its perpetual laboring.

A Traveler's Request for
Shelter in Viterbo

What windows, what rooms decorated for a holiday
        you open
to air, freshness, mild sun. And houses
on every side, incandescent façades,
children, swallows, owls.

The grape carts stop at the main gate in single file;
men and shadows follow each other.

A woman takes water from the fountain,
climbs up the outer stair, looks at
that ship, Viterbo, anchored in the sky,
reenters, disappears inside
the house, the city and time.

New to these streets but not a stranger,
I heard an invalid on the threshold
pray for the destiny of this ark,
its bustle of workers,
rundown houses, animal stock,
shrewd old-timers and its dead.

I left my horses at the gate;
I asked for shelter and indeed begged
to become one of them. Now, you,
take a good look; scan the signs of night.

Las Animas

Fire everywhere, the gentle fire of brushwood,
fire on the walls where a feeble shadow floating
hasn't the strength to imprint itself; fire
further off rises and sinks in loops of thread
downhill across a length of ashes,
fire in flakes from the branches and trellises.

Here not before not later but at the proper time,
now that everything about the festive
and sad valley loses life and fire,
I turn round; I count my dead,
the procession seems longer, trembles
from leaf to leaf as far as the first stump.

Give them peace, eternal peace, carry them
to safety away from the ashes and flames
of this whirlwind that presses
strangulated in the ravines, is lost
on trails, flies uncertainly, vanishes.
Make death what it is, nothing but
death, struggle done with and lifeless.
Give them peace, eternal peace, quiet them.

Down there where the cutting is thicker
they plow, push vats to the springs,
whisper during the hushed mutations
from hour to hour. In a corner
of the garden a puppy stretches himself and dozes.

A fire so gentle is hardly enough,
if enough to illuminate as long as

this undergrowth under life may last. Another,
only another could do the rest
and more; to consume these spoils,
to change them to light, clear and incorruptible.

Requiems from the dead for the living, requiems
for the living and dead in one flame. Poke it:
night is here and overspreading,
stretches its quivering cobweb between the mountains;
soon the eye will no longer serve; what remains
is awareness for ardor or the dark.

On the Threshold of 40

This thought stays with me in this gloomy
village swept by an upland wind
as a cliff swallow diving cuts a fine thread
across the distant mountains.

Almost forty years of anguish,
of boredom, of sudden exhilaration, swift
as in March a gust of wind is swift,
scattering splendor and rain; the delays,
the tearing away, hands outstretched from those
        most dear to me
in my village, the habits of years
suddenly broken that I must now understand . . .
The tree of grief shakes its branches . . .

The years, a swarm at my shoulders,
rise. It was not in vain; this is the work
done by each and all together
the living the dead, to penetrate this dull
world along clear roads and passages
thick with brief encounters and deaths
as from one love to another or a single one
from father to son until it may be crystal clear.

And this said, I can get going
promptly within the eternal co-presence
of the whole in life in death
to vanish in dust or in fire
if fire beyond the flame still endures.

In the Month of June

In the month of June
when the city suspended
high above our confusion
arises to the darting of lights

in the uncertain hour between sleeping and waking
when the body stumbles of its own weight
but wearily gets up again,

in that pause of time between the swallow and the owl
between life and its survival,

since You who shatter the slavery and the pride
—they say—of suffering, come
if you are not already everywhere
clothed as a vagrant,

an invalid, a troubled child.
Follow the timid, approach the lonely.
Tell yourself: virtue without love
is empty.

It is that moment at mid-year
when the homeless shuffles in his rags
on the trampled grass and seeks refuge.
Fireflies glow. A dog barks.

Refugee Camp

In the treacherous weather
a woman climbs up slowly and pulls down
rags strung between poles. A dog whines,
giving substance to shadows.

These are the signs of a stormy day
in the maze of ramparts and ditches;
these are men like herds at rest
or freight held in customs, piled
under tents or in shacks, permanent
or in transit—a sight until dark
of motionless migration without
rest that the just man chosen for expiation,
erect against a doorjamb, observes
between intermittent showers and snowfalls.

The wind carries a splash of dull water.
What are you doing? You are lost in this riddle.
New to this place, one man hesitates uncertain
which road to take; the other, an eel fisher
or sand digger, moves beyond,
firmly pierces this wet cover
cast on the river between thunder and lightning.

The Wolf

When the ice breaks up
animals in anxiety on the floe
stare at the melting seas, the drift of icebergs;

sharks, shuddering, wounded by spears
thrash and perish; salmon,
eager to spawn and dying,
swim against the heavy current

and the wolf
his life-long aching for
forefathers and cubs
crowding his heart

takes to the mountains and is himself again,
nimble on his old paws, alert
to the cry of primeval winds
that blare of love journey plunder.

O life not mine, a grief
I bear from night
and chaos,
you, a sudden twinge, go deep
writhing in pain under the burden.

I live as I can, as he who serves
faithfully, since there is no choice. Everything,
even gloomy animal eternity
that moans in us can become holy. Little
enough—that little cuts as a sword.

To Niki Z. and to Her Country

What voice once heard in laughter and entreaty between island
     and island
and what screech from a swallow darting
between cloud and cloud comes and puts an end
to the lethargy on the shore after years and years of sea.

Who are you? I do not know, but certainly someone like you
     appeared to me elsewhere
in strips of towns seen or lost
behind a veil of rain or under a sky
divided between cloud and radiance.

And silence and uproar of a people fighting on both sides of you.

If here where sea and time
fall in waves
it is not easy to distinguish
echoes from voices—
I am not deceived about the groans
of the unjustly dead. I recognize
the instant the crow narrows its wheeling,
and if air stirs, the hanged man swings.

How lightly you bear this weight!
Suffering for justice lightens
the heart, gives strength and exhilaration
even more in your country, mine too, where
a snake's deceit roughens
the road. Under the clear and shining lamp
everything is full of light and invisible darkness.

It is the season of rain and clearing,
of bewilderment and encounters.
Wave on wave the sea sparkles and moans.

House by House

Gray in April, violet in September
the mountains in flight
carry me away.

This is the village that runs through my mind,
house by house, niche by niche:
year after year its afflictions
and joys were also my thorns,
my wine. This is the time: the separation
of my family, one branch after another, ready
or reluctant for the journey, suspended
between light and shadow an instant,
overturned in passages below, from father to father,
from era to era, that returns alive and bleeding.
The swarm embroils the old bee-hive.

Lives crushed by evils or strong lives,
each with its tribute, offering,
imparting fire to others, falling into
humble routine, until the groping
search for the door
at the end of this dark corridor lasts.
Nothing else. And what remains is man's work
cut off and carried on. Mark,
mark with a steady hand those present, those no longer here,
the living the dead, you who break up and separate.
I observe the plunder but do not despair.

Night and Mind

The night cleanses the mind.

Soon after as indeed you know, we are here—
a row of ghosts along the mountain ledge,
ready to leap, almost in chains.

On the page of the sea someone traces
a sign of life, fixes a point.
Seldom does a seagull appear.

FROM THE HEART

OF THE COUNTRYSIDE

Translations dedicated to Alfredo Rizzardi

Smoke

Smoke rises lazily from the heaps,
shredding from the edges of hovels, from thorns
in strips along the bank in fragments,
spreads in streams through the damp hillside,
shadows a few inches
of tillable land
snatched from the mountain—
it takes the breath away. Quite wet, smoke rises
from roots, tendrils, settles
into the year. There was green fleece
from top to bottom on the stratum; the woodland
crackles branch by branch or silently
unleafs, unleafs; a heavy soaring
of other tongues of fire takes hold of the land
between the heather and broom, searches out the pyre.

"Gianni grows. Is your father still living?"
I look at you as you are, still strong
or just about to fall
on the side of the vanquished,
my dears, my equals, I repeat:
For you death may not be this rotting
of face or shriveled limbs, death
may be something other than an opaque migration
of souls in search of the crater. To the living
a spark of fire in the ash
is much, is enough. Little more than a filament
of life in flesh almost dead
is still life. But if you must
perish and survive
through fire's mediation

it is not this mist
of soaked and burnt
leafy branches that can nourish you; it is a blaze
of transparency lit in other lives.
I struggle; I try that mine may not be unworthy.

Blows

In pruning trees the curved hook
slashes blow on blow. In the cold,
the bright wounds are resplendent.

A time a man in the fullness of years says:
I am what I was in the fellowship of fire
that brightens and gnaws substance: I look after

what burns and becomes ash.
I keep faith in the thoughts I once had.
Really it's no great thing; it's less than little.

Years, yet everything offered me
under the semblance of grief
comes too late to make life true.

Year after year
life follows life
with that fidelity shadow has

while the river flows,
while a spear of grass trembles
between the blades of the mower

and man hardly escaped from the ordeal
whole or deprived of his bounty,
raises his head for the next blow.

Bird Hunt

What a livid sea rushing against the plumbline walls
        of the bunkers!
What flocks of birds waited for to go by or return,
screeching more than any other time: "It is autumn,
the time of your birth in this life" now when one by one
they fall under buckshot, glide in the wind
in falling, and as far as the eye sees, the woods
shed leaves, shreds of fire branch by branch,
fragmented life still breathing under the feathers.

This instant, here where a dog flushes a partridge
and sometimes gypsy kings encamp
for a brief stay, a few hours,
in moving from hamlet to hamlet, leaves
and birds native and migratory,
slight and heavy, fall to the soaked
if not yet cold soil; time
of my birth, time and place together,
remembering my murdered dead,
my fallen ones under fire; shortly
before, my forefathers; then my brothers—
the wind of life, one and the same
as plunder and death,
strikes me square in the face, cuts my breath short
as I raise my hands to these trees, and still eager
for supper, pluck fruit.

"It is the time of your birth." They sleep;
a life to die in, they perish
in due time. Joy and darkness spread
through dead leaves, through leaden wings
to conquer and atone for all that has an end.

The Strong Filament

"Pass by our house some day.
Think of the time we were all together.
But don't stop for too long a while."
It is the voice of one of the faithful,
who, summoned, prepared for her passing.
She wept as she fixed the last meal,
then listened to the bare crude judgment
as it was spoken, that voice hardly deeper in its tremor,
hardly more touching now that it comes
from the frontier of shadow and rips
as it can the curtain of years and pierces
the coverlet of toil and degradation,
searches for the wind's way, gives itself to it
until the wind leaves it to itself, stays around
as a guest where it belonged, afraid
and lost in these first dawns of the year.

The time is that harsh hour, day hardly begun,
when cold reveals the city
livid to its stones,
its corners cutting. Indoors in shadow
milk is poured into cups, bread
toasted; half-awake, a boy slurps
as he jots down a new day in his diary.

In that clotted heat that is more hers,
in that bubble of life that is more tender
for her brought up to be patient on poor
humble land, I listen to her faint voice
reaching to those still heavier ones,
still dulled from long sleep, asking for

refuge, trying to mingle.
I say: peace and silence be yours. I say . . .

To hear voices, gone by, snares
the just, flatters the too weak,
the too human part of love. Only the
word in unison with the living
and dead, the live communion
of time and eternity is worth cutting off
the strong filament of elegy.
It's hard. Everything else is too tough.

"Pass by our house some day.
Think of the time we were all together.
But don't stop for too long a while."

The Fortress

Haze in mid-air
on an October afternoon
muffles the few voices, slackens
the descent of doves
to the gun carriage,
cushions the sound of footsteps on the gravel.
Here a man gripped in the vise
of old age sits
a long time, unwillingly turns his face,
sizes up from head to foot someone coming up
the grassy-edged path to the bench,
then shuts himself into his dullness.

His is the face of eternal old age
cut into this dry and wasteful time.

Much water poured
on the fire
of youth and sociability takes away
the warmth of life in this life.
Life, if it lasts, is no more than the slow
extrication of souls, one from another
at the exit of the maze. There happens
what happens in spring or summer
when the wood is shaken,
when in an instant
the darkening of the sky on hikers
dampens their gladness and one by one
they withdraw yet remain together
within the prison of their selves, their true hell.

You to whom at the brink of day
I say so many times, "Help me,
guide my soul with your advice"
if the silent wisdom
of the dead at one with God could speak,
I know what you would say,
You would say, "Put your strength
and your toleration of man to the test."

What you ask of me is not little.
But if the price is this
for complete knowledge
and for deeper atonement
I shall pay what is due,
now, at the moment,
in the contemporaneity of all time.
                              The old man
scatters crumbs from his threadbare pocket.
Two, three doves come directly and roost.

Within the Year

In a great hurry, motor switched on,
two words of goodbye, two hackneyed phrases
about days past, those fallen
behind, not for this less alive
than the living, "since childhood is immortal."
I get off and already you take the curve and step on it.

This is the time between lunch and supper
when under my windows boys plan what to do
in what's left of a Sunday.
The ice-cream vendor stacks his cones,
pushes his tricycle, honks his horn;
the sun comes out, goes down under clouds;
it's drizzling a little; the sprinkled dust
has a strong smell; jasmine's
stronger; grain is tall as it must be
at this time of year. Everything's right;
figures add up to a total, precise
to a fraction: or at least so it seems,
at least so the heart is prepared to believe.

Shots

The first gunshot skims
the blossoming rosemary and broom.
The hunting season in these places high above the sea
tears into venerable sleep, distracts the custodian
      of the ruins,
mows down the flock year after year.

Get used to life less full
with not enough light and heat;
cling to bitterness, mindful of it, I declare.

Resist, hold fast the thread still taut
between the gradations of eternal fever;
smile, finish your work.

While stroke on stroke
the grandfather's clock of the years
scans the time of a goodbye,

while a hawk, wings fixed, soars
in flight; while a dog points,
and a bird, flushed, takes off and rises
full tilt against the draw of the shouldered rifle;

so man trembles
and as he meets his suffering head on,
he writhes between cowardice and courage.

The Country Bus

The country bus roars as it moves jerkily.
A native here recognizes the ridge
as he follows the entire blue cavalcade
crest by crest in the distance; the wind
profiles the first mountains,
their summits scorched; it makes
the coloring livid,
more ashen than flame
the oakgrove wears in winter
on this plateau;
it slashes, handicaps the mules on the slope,
shrieks over heaps of embers. The others:
someone repeats the breviary in a low voice,
another dozes; a third talks of his deals
in oxen, wool, grain, and turns,
if he does, one blank eye to the window.

Passengers on a trip, we sit here,
some anxious to arrive, some
thinking of their departure, some perplexed.
The shepherd moves his flock into line,
nudges them close to the shoulder of the rise,
unblocks the road; the bus moves ahead
bouncing heads and scruffs.

On this strip of my homeland,
I shut and open my eyes. Stiff against the back,
I listen to these folks, this wind;
I more alive through neighborly mediation
than by my own flesh and bones.

Somewhere

One who stumbles on his clubfoot
strews his venomous speech
among knots of street-talkers. It is the time
when the fire of revolt
kindles those resigned to half-a-life. An old man
with a good deal of wisdom then shakes
his head, says: "That's how things are," disappears
behind a curtain of reeds. People
run, a deafening shot, some
scream. The square returns to what it always was.
A day, a day of anger in a village
on this dry land, scratched
by a plow, hardly more than a nail.
We, two, three witnesses here by chance.

That little of the world that appears
in this slot of the senses,
in this crack of the mind,
hones the knife of judgment,
makes its firmness cruel.
What are you doing? Your hand raised, you dare
     to brandish it?
I stir my own faithful thoughts
with thoughts hatched in the nest of others,
as fledglings of the shrike. I struggle
so as not to judge with an iron heart,
so as not to cut to pieces what is whole.

From the Tower

Its humps licked by the wind, this gray earth
in its gallop to the sea,
its crowded herd under the ridge
on the sweep of scarps seen
in the dizziness of barriers, spins
light, spins mysterious light years,
spins a unique destiny in many ways,
says, "Look at me. I am your star."
And at that instant the thorn of life
pierces the heart more deeply.
This Tuscan earth, harsh and dry,
over which run the thoughts of those who remain
or of those reared by her who leave!

All my more than forty years swarm
outside the bee-hive. Here they look
more than elsewhere for sustenance; they ask
about us, of you interred in the crust
of this luminous planet, and it continues,
continues by death teeming and life
tender, hostile, bright, unknowable.

From this lookout tower much catches the eye.

How Much Life!

"How much life!" A child raises his high-pitched voice
where birds and birds snatched chirping from branch to branch
spin among the wastage of woodland leaves in the cold counterlight
and trace a screeching track of feathers, leaving behind them
          those broken phrases
of speech arrived at the crux; a joyous
escape while men in ambush
prepare their extinction. "How much
life!" repeat those last luminous flutterings of wings
throughout the woodland between the sea and marsh.

And here, in places fairly distant,
but in a pitiless time, as
I cross this street of banking houses,
friends of another day,
no sign between us,
are dragged by a dark wind through the guarded doors.
I see them worried like the birds, late-comers overcome
and burnt by indefinable fire,
wasted, not yet spent, a presumption of power
where there is no power, pride
in a faith where there is no faith. "How much
life?" that nine-year-old voice repeats
to his quite grown-up, quite
clear conscience, "How much life!"
No one ever perceives of life
as strong as in its doom.

Senior

        To the old
everything is too much.
A teardrop in the crack
of a rock can slake
thirst, a small thirst. The end
and waiting for the end ask
little, speak softly.
But we in the fullness of age,
in the furnace of time, we? Think about it.

IN THE MAGMA

Translations dedicated to Luciano Rebay

. . . nisi quod pede certo
differt sermoni, sermo merus . . .

HORACE, DISCOURSES *1, 4*

. . . were it not for strict meter,
it would be merely prose.

Near the Bisenzio

The frozen mist blackens the millpond of the tannery
and the path that follows the bank. Four come out,
idlers by their walk, idlers even in stopping face to face,
I know not whether seen or never seen before.
One, the most overwrought by frenzy, the laziest,
confronts me, says: "You? You are not one of us.
You were not burnt as we by the fire of struggle
when it blazed and good and evil glowed on the pyre."
I concentrate on him without answering his feeble weak eyes
and catch an inquietude quiver along his lower lip.
"That was but one time to redeem oneself"—(here a trembling
twists into a convulsive tic)—"or to lose oneself and that was it."
The others constrained by an unpredictable pause
give signs of annoyance but do not speak;
they shift their feet in cadence against the cold
and chew gum looking at me or no one.
"Are you dumb?" The tormented lips curse
while he pushes himself forward and withdraws
often frenetic until he is there,
stopped between irony and fury, and leaning against a post
looks at me. And he waits. The place,
that little that is visible, is deserted;
the mist presses close to the bodies
and allows the soaked earth of the embankment to appear,
and *il cigaro*, the fat plant of the ditches, that trickles mucus.
And I: "It is difficult to explain to you. But you know the way
for me was longer than for you
and passed through other sides." "What sides?"
As I do not go ahead
he stares at me for a time and waits. "What sides?"
My friends: one rocks himself, one springs his body on his heels

and all chew gum and look at me, me or emptiness.
"It is difficult, difficult to explain to you."
There is a long silence
while everything is still,
while the water in the millpond whispers.
Then they leave me and I follow them at a distance.

But one of them, the youngest, it seems to me, and the most
      doubtful,
goes to one side, loiters on the grassy edge and waits for me
while I slowly follow them swallowed up by the mist. Aligned
by now, but without my stopping, we look at each other,
then lowering his eyes, his smile is an invalid's.
"O Mario," he says and steps to my side
along that street that is not a street
but a twisting trace lost in the mud.
"Look, look around you. While you think
and reconcile the dials of the mind's watch
on the movement of the planets for an eternal present
that is not ours, that is not here or now,
turn around and look at what became of the world,
think of what this time asks of you,
not profundity, not daring
but the repetition of words,
the mimicry without the why or how
of gestures in which our crowd lets loose
bitten by the tarantula of life, and that's it.
You say aim high beyond appearances
and you do not feel it is too much. Too much, I understand,
for us who are after all your friends,
young men worn out by the struggle and more than by the
      struggle, by its humiliating deficiency."
I hear my friends' footsteps eclipse in the mist
and this voice come by snatches and broken gasping.
I answer: "I work even for you, for your sake."
He is silent a while hardly receiving this stone in exchange
for the grievous sack, emptied and scattered at my feet.
And as I say nothing else, he again: "O Mario,

how sad it is to be hostile, to tell you we refuse salvation,
nor eat the food you proffer, to tell you it offends us."
I let his breathing calm down little by little cut short by
        breathlessness
while their footsteps die away
and only the water of the millpond whispers from time to time.
"It is sad but it is our destiny: to live together at the same time
        and place
and make war for love's sake. I understand your anxiety,
but I am the one who pays the entire debt. I accepted this fate."
And he, now bewildered and indignant: "You? Only you?"
But then he cuts off the outburst, grasps my hand jerkily
and shakes his head: "O Mario, but it's terrible, terrible you are
        not one of us."
And he cries and even I could cry
if it were not that I must show myself a man to him who has not
        seen many.
Then he runs away sucked up by the mist on the path.

I stop to gauge the little said,
the much heard, while the water of the millpond whispers,
while high wires buzz in the mist above poles and antennae.
"You cannot judge these years lived with a hard heart,"
I tell myself, "others will be able to in a different time.
O pray their souls may be stripped
and their pity be truly perfect."

Through the Clinics

In the dusk of verdant cities,
as this is, with few houses, unfortunate in hospitals,
fireflies pierce the insidious darkness, flash lightning
sharper-edged than elsewhere, more mercilessly lost in the uneasy
     eyes
of the bedridden. "Where did you live?
Where? I looked for you far-off no less
than the swallow when it soars
to explore the white regions of rain
and meanwhile mites overrun the nest."
Someone murmurs to himself
while fireflies swoop down and wing up
the black abysses, the pits
known darkly to the senses,
while the butterfly of the blood roused from lethargy
dances, swarming up on wavering thoughts.

"I was not far away,
I was within you
when you went out for alms,"
answers some ghost
from the burrows of night one knows not whether mocking or
     crying,
and meanwhile from their treacherous quiet,
resigned to nothing, years of obscurity and passion rise,
crucifying body and soul between regret and agitation.

Here near these clinics
(the moment when fireflies pick
at the cluster of grapes of the night and signs
of a felicity never reached

or escaped as they pursue themselves, they become scalding
for the weak, for the sick),
I join with every unstable power
of resentment and unfulfillment; I touch
the unsociable element that keeps the world hanging.

The mind hurries to distinguish between sickness and health
sickness and health, it repeats
until the parts of this knowledge possessed
intermittently in the dusk fit perfectly.

Ménage

I see her again now no longer alone, different
in the innermost part of her house,
in harmonious light, without color or time, filtered by curtains,
her legs drawn up on the divan, crouched
near the record-player kept low.
"Not in this life, in another," flashes her joyous glance,
yet more evasive and as offended
by the presence of the man who curbs and crushes her.
"Not in this life, in another." In the depth of her eyes I read this
        well.
She is a woman not only to think of it; she is proudly certain.
And this is not her ultimate charm
at a time like ours that really is not extraneous or adverse to her.

"I think you know my husband." And he pouts an importunate
        smile,
as quick as transient, almost a desire to throw her off his back
and drive her behind, beyond a wall of mist and years;
and as he comes up to me, he has the air of one who gets
to the point face to face with men.
"Is there something to dig out of dreams?" he asks me, fixing his
        blank empty eyes
upon me, I know not whether of a monster in some *villa triste*,
        or of a guru.
"Something of what sort?" and I look at her radiating tenderness
toward me from her golden, witty and liquid glance
that is a little sorry for me, I believe, in being in her clutches.
"The dreams of a soul mature to accept the divine
are dreams that create light, but at a lower level
are unworthy, an expression of the animal, and that is it," he adds
and sets his impenetrable eyes I don't know if they look or
        where.

Again, I do not understand if he questions me
or goes on talking with no beginning or end,
and not whether he speaks with pride
or something dark and inconsolable weeps within him.
"But why speak of dreams?" I think
and search through my mind for a haven
in her who is here, present at this moment of the world.
"And aren't you making up a dream?" he continues while from
        the street there leaps
a strident scream of children chilling the blood.
"Perhaps the limit between reality and dream . . . " I murmur
and listen to the sapphire needle
in the final grooves without notes but a click.
"Not in this life, in another." She exults more than ever,
an unbearable light emerging
from her look, proudly displaying other thoughts
than the man's, whose caresses and yoke she bears and perhaps
        desires.

The Office

I see him hardly beyond the threshold, standing at his post
bent over his desk, indifferent
to the diminished fever that agitates
that light in an aquarium or false temple,
and I am mysteriously repelled and attracted.

Meanwhile in his tiredness he lifts up his face
and in that face the emptied and white glance of an invalid or
      idiot.
I can't place him but by a sudden twinge I know he is not alien
to my past and as he stares at me
I try to find him not among friends
but in the stifled and inexplicable rancors of a purer age.

"Why are you here?" he asks me,
stressing the words more than deserved,
or maybe it is I, already too bitter and bristling.
Perhaps it is the gibberish of a disheartened man, lymph squeezed
      dry,
and it is enough for me to revive
the impalpable grudge that was between us in another time a little
      after childhood.

In that sequence of shelves and paper, I look at him without
      replying
and ask myself if that is his realm
or the prison that has discouraged and wasted him.
"How read fate in so inexpressive a face!",
I say to myself while my malice weakens at once,
yet my desire is he talk to me again, even at length.

Then I am silent before him who waits,
I waiting too, and meanwhile I think
if there may be in this confrontation
something not due to chance only
for a debt to be annulled in an age still living
no matter how past, or because an obscure end must come true.

"The girl fell into your power. Of what I know
she took no glory in it." His whimpering
and distracted voice is hail on my face
not without strength to hurt, animating
that mask with a sneer or smile even worse than mourning.

"Oh, it did not happen as you believe," I answer
and then I see again the where and the when
and in a precise corner his figure already moth-eaten.
I don't think of defending myself; I think of the closed-in
knot of that suffering left shut
and locked in a moment of his life without redemption.

"I know your kind. You sacrifice
yourself and those nearest you, blinded by your presumption of
    art.
That which is lost sometimes does not even pass through your
    mind."
After a while he continues: "It was my salvation, also her . . . "
And he sharpens his glance that much
the whiteness of his eyes,
fixed on me, lights up.
"Who can say that?" I dare, not finding another word
to join us to the obscure sense
of good and evil, received and accomplished.

But he is not a man to come to me
on this point that should unite us
as friends expert in the grief of the world.
I see him shut in his own hurt, his teeth clenched

and I know not if he recriminates or if he broods
in such a way his strength challenges his hell.

The silence that follows in the room
where we are not alone, and yet lonely,
is an enormous silence without confines or time
while the blade of the fan hums
and wheels with a trembling of detached papers
and I think of the struggle for life in the sea-depths and of plankton.

"I am not yet a failure." Eyes goggling,
he then explodes
breathing in my face the breath of tobacco and alcohol.
"No more than me, no more than anyone else,"
I murmur, sucked up by his passion,
and I look beyond the windows at the throng
into which in a little while I will disappear.

THE WHIRLPOOL OF

SICKNESS AND HEALTH

Translations dedicated to William Jay Smith

. . . carrying to the light that which lives,
waking someone who was dead . . .

RIG-VEDA

From *The Whirlpool of Sickness and Health*

2

At the other end of the line anguish was long still.
Her silence came clear and strong
in the city's roar over the receiver off the hook.
What can I do? her still adolescent old boss
asked himself
overtaken by his uncluttered prescience
in his office on the 60th floor at the top of the pyramid.
Too late to hang up,
useless to dial the number again.
He was troubled, his ear glued to a funnel of silence,
to that uncovered cavern of the world's sorrow.
In turn, he kept quiet when I came in.

"Is there a future for man?" Reporters
never fail to ask him. The sly canvassers
of the problem, sociologists, ideologists,
busybody priests impatient with the Word are insistent.
He, ready, indefinitely futurable, always
fobs off an exhaustive non-answer,
surely not for them but for the microphone—the only
innocent as far as I know among those sinners against essence.
Besides, his pretentious jacket,
his talkative wife,
his glory set up
economizing mostly on his soul answer for him.

And then away, suddenly vaporized—
and behind them, the undertow

of a sterile and inexplosive age
that moreover writhes in pain
and hastens the time of paralysis and coma.
They go a slight distance, I am sure, but go
in melted water, clouded,
in the cold of a black springtime
open to a few emeraldine fans beyond the terminal.

3

Happiness in the past. The future of stellar conjunction—
she who moves forward in time, twists herself into her loneliness
or perhaps into a net full of broken meshes, encounters,
new ties, new fade-outs
while the signs of her power last unbroken—this
and something else that's mumbled of as "life"—

I think of when I think of her who was queen of this place
as I observe under the vertical sun
the dazzling and inexplicable array of her works,
the pigeons briefly slapping their wings in a frenzy
along all the palace walls at the entrance to their shelters;
then, between stone and air no other motion in that box of fire.

4

Changing, unchanging?

                                             Master-teacher

of that slalom
or no, really
knowledgeable to a higher degree,
her answers are but smiles.
Neither do I prolong the doubt or inquiry;
I look at her radiant eyes and think
of the reflection in a crystal ball
(what clairvoyant have I ever consulted?)
with its sign that deciphers all inner signs;
outside, mirrored by the convex surface,

I am unaware of fermenting putrescence:
souls in torment for not attaining spirit
or only the adder of misery
about the bright red or faded great walls of Islam.
I look at her eyes, not really eyes,
those meshes of light between sense and sense no longer, vibrating,
    liquid.

❖ ❖ ❖

While she talks of the towers of the Parsees
something like a flight of vultures
crosses her eyes, not darkening them;
she flares up as if caught red-handed
in her proud counterblow of ecstasy
and waits for my ironic answer;
instead, I don't interpose a word;
drained of thoughts I think, not of her,
but of predatory carrion
only in part deciphered by sense.

6

Or when your eyelids are inflamed
by the brutal appearance of history far different from its image—
indeed not the dull
commutation of things in conflict with the possible
or rather with the illusory you think of in your room
(how deducing God knows!) yet the obscene palsy
of a time that skids on itself
with motions of anger, an age
crazily unnailed from its sloth
bites not the still unhardened matrix of its present
and burns strength, spreads a murky cloud studded with fiery
    stones;
each one of these falls and pierces you well under the skin.

"I know you'd like to escape it"

a half-mocking face, not momentary,
perpetual as life, a sister's
not by blood, espies you.
But where? Under the thatches of hair the obsidian
points of two young eyes smile,
older and younger than you, two fires
imperceptibly cross-eyed, two souls
of the world at war yet attuned to each other.

✧   ✧   ✧

At the first brightening after the rain
the standard-bearers, the ladies, the horsemen,
the city of metallic flakes, sparkling,
she, memory and song behind her,
crosses my name from her guest list,
arranges for the transfer of my luggage,
gestures my well-known distant destination.
I no more than greet her
or look in her face for the past
of this very instant, even less
for the flash of mutual awareness
ab antiquo. I pay up and leave.

7

"The dissimilar, different
entirely from me—you have experienced it."
That part of her soul, a child's or a puppy's,
sometimes whispers in her sleep
offended by betrayal and darkens,
hearing herself still searched for,
"but not for the sake of fruitful love—she regrets—
only to preserve its memory and cold ecstasy together."

—I have been no more unfaithful to you
than an insect liberated from its chrysalis—
I wish I could answer her without lying,

and it would have been enchanting to think of her
blinking her eyelids
perhaps slightly dazzled by my firefly sparks
there in the womb of darkness that swaddles us.

"The way to growth
is not so smooth,"
I tell her however and let her melt
light and tears into turquoise bubbles.
"The grindstone of the world resembles you but little."
Imperfect knowledge
even this I don't deny:
through the mirror and in the image,
was said one time
which is in truth no time—signs
perhaps, maybe ghosts, yet always knowledge.

NOTES

Notes

(Notes by M.L.; comments in brackets by I.L.S.)

IN THE DARK BODY OF METAMORPHOSIS

Part 1. *the defeated face/of a master thinker of the western world:* The crisis of Western philosophy, the break between thought and life, is symbolized in the figure of a philosopher whose identification doesn't matter. [In a letter, Luzi named Russell and Adorno (Theodor Wiesengrund) as alternatives. During our final sessions, he said he was thinking of any "rationalist" . . . "any philosopher who relies on reason."]

the drowned city: Florence flooded by the Arno (November 4, 1966) is also the symbol of the ruin of the "city," the modern "polis."

with a bewildered glance/a figure surfaces: This is the figure of the woman-mother, an epiphany or manifestation of feminine polarity, present in various aspects throughout the entire poem.

fish-swarmed city: The city as an aquarium, men as fishes who pass by each other without saying a word. [There is a silence, hollow and awesome, as in Dante's *Inferno* 1, 22–24: "And as that man breathless and gasping escapes from the sea to the shore, he turns to the perilous water and stares."]

Part 2. *the lost revolution:* Disillusion, decay, and corruption of faith in civilization and its institutions.

many good habits, trips, the arts: Unreal life with its comfortable, if sad, substitutes.

Part 4. *Esenin:* [The anarchist poet, married to Isadora Duncan.]

Part 5. *Church, Church:* The title and beginning of a satirical popular anti-church song written by a deist, "who expresses delusion (and naturally, love for the Church). The entire section foreshadows feelings of criticism yet of piety for the Church that I represent, even as a woman, as you see, from her changing point of view." [This, from a letter written January 2, 1970.]

Something imminent dominates her: The Church is confused and weeping.

"I almost don't remember him": The woman who has lost her first husband in a Nazi concentration camp feels a dark visceral grief. Analogically, the Roman hierarchy has divorced Christ and so moves no deep sympathy in this woman, whose dull blind suffering has isolated her from Mother Church.

the Shepherd of her anguish: The reference is to the liberating image of Pope John XXIII.

Part 6. *the queen of the rocky city:* Tamara, the legendary Georgian queen. [The entire section suggests the illusory happiness men place in the past. They are dazzled by certain splendid ruins. What is left man but the drama of history and life itself, seductive and unfailing. Woman as a creature, pliant to mutation, available for hope, trusting in the future, the trustee of the very continuation of life.]

Part 7. Life as vital reciprocity where the mutable and the eternal merge.

PROPER PERSPECTIVES

Las Animas: Jorge Guillén told me that in Spain this is called the Day of the Dead.

To Niki Z. and to Her Country: The reference is to the horrible struggle in Cyprus.

IN THE MAGMA

Ménage: [During World War II, any house in occupied Italy where Nazis tortured their victims was called *villa triste*.]

THE WHIRLPOOL OF SICKNESS
AND HEALTH

Part 4. *Parsees:* [Zoroastrian fire-worshipers who remained in Persia after the Moslem conquest. They exposed the bodies of the dead on structures called "towers of silence" that they might be dissipated without polluting the earth.]

Part 7. *through the mirror and in the image:* 1 Corinthians 13:12.

110